George Cary Comstock

An Elementary Treatise upon the Method of Least Squares

With Numerical Examples of its Applications

George Cary Comstock

An Elementary Treatise upon the Method of Least Squares
With Numerical Examples of its Applications

ISBN/EAN: 9783337279530

Printed in Europe, USA, Canada, Australia, Japan

Cover: Foto ©berggeist007 / pixelio.de

More available books at **www.hansebooks.com**

De Fidiculis Opuscula.
Opusculum VII.

THE

Arts and Crafts Book

OF THE

WORSHIPFUL GUILD OF VIOLIN-MAKERS
OF MARKNEUKIRCHEN,

From the year 1677 to the year 1772.

EXTRACTED AND ANALYSED BY

Dr. Richard Petong.

TRANSLATED AND EDITED BY

Edward and Marianna Heron-Allen.

LONDON:
PUBLISHED FOR THE AUTHOR BY H. S. NICHOLS AND CO.,
3, SOHO SQUARE, W.
1894.

PRICE ONE SHILLING.

Dedication.

To Dr. Joseph Joachim.

Dear Dr. JOACHIM,

During the fifty years that England has been, as you felicitously expressed it, your "second home," our countrymen have learnt what Germany has done for the art of Violin-playing. It is, therefore, with a keen sense of the fitness of things that we address to you these observations upon the services rendered to the art of Violin-making by your countrymen.

We are, most sincerely yours,

Edward and Marianna Heron-Allen.

London,
    July, 1894.

J. W. WAKEHAM, PRINTER, 4, BEDFORD TERRACE, KENSINGTON, W.

1894.

THE ARTS AND CRAFTS BOOK

OF THE

WORSHIPFUL GUILD OF VIOLIN-MAKERS

OF

MARKNEUKIRCHEN,

From the year 1677 to the year 1772.

EXTRACTED AND ANALYSED BY

DR. RICHARD PETONG.

TRANSLATED AND EDITED BY

EDWARD AND MARIANNA HERON-ALLEN.

––––––

This guild which celebrated its 200th anniversary in the year 1877, undoubtedly laid the corner stone of that branch of industry to which Markneukirchen owes its world-wide renown, its distinctive character, and its present prosperity. Among its archives may be found a book bearing the above title, the contents of which might, from many points of view, be incentive to a fresh departure in the manufacture of musical instruments in this locality, and might suggest a wider scope for the operations of this branch of industry so long established in the Vogtland.

How, in the early days of the history of this industry, a mere handful of independent masters and workmen applied themselves to the manufacture of violins and 'cellos, and how this artistic handicraft (its practitioners goaded to emulation by the superior models brought from foreign lands by masters of the art) became gradually extended to the manufacture of many other kinds of musical instruments is now well known. Especially did this become apparent when the manufacture of harps attained such proportions as to call for the formation of an independently organised guild.

We find in the "Crasseltschen Chronique" —a work so scarce as to be almost unknown --that, according to contemporary tradition, the population engaged in the manufacture of musical instruments at the commencement of the 17th century appears to have been composed of the descendants of a small number of Bohemian and other immigrants, who had left their distant fatherlands in order that, in a new country, they might enjoy liberty to profess and practise their Evangelical religion, and had settled in Markneukirchen, bringing with them the arts and crafts of their native lands. To this nucleus were added in all probability other families from neighbouring localities, but the number of parent, or head families, was comparatively only slightly increased. In the year 1627 they numbered probably no more than 66, this being the total number of houses occupied by fully enfranchised citizens at that date, the privilege of free citizenship being conferred by the High Court of Justice through the Kurfürst Johann of Saxony.

We may take it as certain that at that time there were already a certain number of brothers and other relatives of some of the principal inhabitants, who lived apart from them in houses of their own and on terms of equality with them as regards their rights of citizenship, and the number of less well-to-do inhabitants (those not possessing the same municipal privileges) must have been inconsiderable. Those colonists (if we may so call them) who joined the community later on, stepped into the places of such freemen as had died, or had continued their wanderings farther. Hardly one-third, however, of the inhabitants possessing full rights of citizenship can have devoted themselves to the craft of violin-making, and the houses that are still engaged in the art and are flourishing at the present day can, as a rule, trace their history back over a hundred years. The following complete chronological record is a key to the list of the original masters of the art, as well as to those subsequently admitted. The book begins with the words : " In the name of the Holy Trinity, Amen ! " and proceeds to give the 12 following names as those of the exiles from Graslitz who, as " Fundatores " or Pioneers settled in Markneukirchen under the direction and inspiration of God. In order to facilitate a rapid survey the register number of each family is added in Roman numerals.

1. I.   Christian Reichel.
2. II.   Caspar Schönfelder.
3. I.   Johann Caspar Reichel.
4. III. Johann Georg Poller.
5. IV. Caspar Hopf from Klingenthal.
6. II.   Johann Schönfelder.
7. V.   Johann Gottfried Böpel.
8. VI. Johann Adam Kurzendörffer.

9. VII. Johann Adam Bopel.

10. VIII. Johann Georg Schönfeld.

11. IX. David Rudest.

12. VIII. Simon Schönfeld (at that time the last master received into the Guild).

Of these, the two Reichels, Hans Georg Poller, Caspar Schönfelder, and Caspar Hopf officiated as head masters of the Guild until 1690, during which period they admitted the following masters and skilled workmen :—

13. X. Andreas Hütel, an exile from Graslitz, who had migrated thence to Markneukirchen, 1678. New citizens, on being admitted to the Guild, enjoyed the privilege of paying only half the municipal tax, viz., eight florins, whereas the entire tax, amounting to sixteen florins, was exacted from citizens who were not members of the Guild.

14. XI. Hans Georg Dörfel, an exile from Schöneck, admitted in 1678.

15. XII. Nikol Kolb, born at Schöneck, admitted in 1678.

16. XIII. Hans George Ludwig, from Klingenthal, 1680.

17. XIV. Hans Friedrich Dörffler, from Klingenthal, 1680.

18. I. Georg Reichel, son of Christian Reichel, 1682.

19. XV. Wolff Nikol Lippold, violin maker, from Klingenthal, 1683.

20. Sebastian Dörffler, from Klingenthal, 1688.

21. XVI. Simon Pöhlmann, burger and " Defensioner," of Markneukirchen, 1696.

Until the year 1704, when Johann Caspar Reichel is described as head master of the Guild, the names of the heads of the Guild are wanting on the admittance protocols; there were admitted meanwhile as master-workmen :—

22. Christian Reichel, son of Johann Caspar (3), and

23. Johann Schönfeld, son of Simon the journeyman (12), both admitted on May 3rd, 1697.

24. Johann Reichel, son of the deceased Christian (I.), 1677.

25. Johann George Dörffler, son of the deceased Sebastian (20), 1677.

26. XVII. Adam Braun, apprentice to a violin maker of Markneukirchen, but not son of a master, and therefore liable to a tax of 9 florins on admission to mastership, 3fl. 9gr. for the two years spent as a journeyman, and a further 3fl. to be paid to the exchequer of the Guild; whereas at the admission of sons and masters a half-tax of five florins was deemed sufficient, and though all members of the Guild had to contribute to the expenses of the banquet held on the admission of each new master-workman, in the early days of the Guild, personal circumstances were taken into consideration, and a reduced fee accepted in lieu of the full contribution.

27. XVIII. Adam Vogt, 1699.

28. Johann Michael Hopf, 1702, Nov.25th.

29. Georg Caspar Hopf, 1702, Nov. 25th.

30. XIX. Johann Dengel, who married the widow of Johann Friedrich Dorfller (17).

31. Conrad Adam Schonfelder, a master's son, 1704.

32. Hans Georg Kurzendörffer (see 8), 1704, July 14th.

33. Hans Martin Schönfelder, son of Master Caspar (2), 1704, Nov. 29th. Johann Reichel going bail for the five gulden due on taking up the mastership.

34. Christian Friedrich Dörffel, 1704, on the same day.

35. XX. Christoph Adam Richter, violin-maker's apprentice, not a master's son, 1708.

36. Johann Caspar Reichel (junior), son of Johann Caspar (3), 1708.

37. Johann Reichel (junior), eldest son of master Georg Reichel, 1708.

At the last last three admissions, Johann Caspar Reichel having been killed by a furnace explosion in 1706, Caspar Hopf figures as Headmaster. Afterwards, from 1709 till 1712, Georg Richel and Simon and Caspar Schönfelder occupied the post. In addition to the above were admitted :—

38. XXI. Johann Adam Lorenz, violin-maker's apprentice, son of a citizen, who had married a master's daughter, admitted in 1709.

39. XXII. Johann Klemm, a worthy citizen, who had learned lathe-turning from his father, admitted in 1710.

40. Johann Georg Dörffler, son of Sebastian (20), 1710.

41. Georg Adam Reichel, son of the deceased Head Master Johann Caspar, 1712.

42. XXIII. Johann Gütter, 1712.

43. XXIV. Elias Pfretschner, who had to pay 21 thaler for the rights of mastership, 1713, when Georg Ludwig (16) was Head Master.

44. Peter Reichel, a master's son, who only paid 6 thaler as a half contribution to the cost of the admission banquet, 1713.

45. Johann Friedrick Popel, youngest son of Johann Gottfried Popel, 1715.

If the later way of spelling Popel with a B be incorrect, we must reckon one family less—the same might be said of Schönfeld and "Schönfelder."

The spelling of "Reichel" is also uncertain and variable.

46. Johann Adam Pfretschner, son of Elias (43), deceased, 1716.

At that time, and also in 1729, the two cousins, Georg and Christian Reichel, were Head Masters; Christian Reichel, is alluded to by that title as late as 1732, and with him Simon Schonfeld, Johann Reichel, and Conrad Adam Schonfelder.

In due course the following attained (8) the rights of mastership.

47. XXV. Johann Adam Jaeger, who, being an orphan, could not pay the full expenses of the admission banquet, and therefore only paid 16 thaler in consideration of his poverty.

48 Johann Georg Klemm, son of Johann The Turner (39), as Peg-turner, who paid 9 thaler in satisfaction of the full dues, and the Master Banquet, and served his apprenticeship like the others, 1719.

49. Johann Georg Reichel, son of Georg Reichel, the Head Master, 1722.

50. XXVI. Johann Caspar Hicker, violin-maker's apprentice, who paid 23 thaler for the Mastership Banquet, 9 thaler as compensation for his Diploma masterpiece being unfinished, and 3 thaler for his fees as a journeyman, 1722.

51. XXVII. Georg Carl Kretschmann, violin-maker's apprentice, 1723.

52.   XXVIII.   Wolff Conrad Kessler, violin-maker's apprentice, 1723.

53.   XXIX.   Johann Michael Seidel, 1723.

54.   XXX.   Johann Adam Glier, 1723.

This family was settled in Markneukirchen as early as 1560; one Andreas Glier (who was schoolmaster in 1587) is described as a native (Stadtkind).

55.   Johann Reichel, son of the Master Johann Reichel the elder, admitted in 1724, whose dues to the amount of 6 thaler and 18 groschen were immediately paid by his father.

56.   XXXI.   Johann Andreas Hamm, son of the Klingenthal tailor and lawyer, Andreas Hamm, who learnt the art of violin-making in 1724.   As he could not provide the Mastership Banquet as a citizen, he paid for everything 35 thaler, which was the sum exacted from strangers.

As the violin-making industry in Klingenthal complained of the heavy fees payable to the Guild, which threatened to prevent the appearance of any more postulants for the estate of master-workman, these fees were lowered to 6 thaler, which was the sum payable by the sons of native master-workmen. But as Hamm again solicited admittance in the following year, wishing to pursue his art in Markneukirchen instead of in Klingenthal, the dues were lowered to 34 thaler in consideration of the six already paid.   (The meaning is here hopelessly obscure.—E. H.-A.)

57.   XXXII.   Johann Georg Hamig, violin-maker's apprentice, 1725, paying 16 thaler in dues.

58.   Georg Simon Schönfelder, a master's son, paying six thaler in dues, 1727.

Within the first 50 years of its existence only 46 men joined the Guild as new masters. As probably few attained the rights of Mastership at 20 years of age, and the founders of houses were mostly of riper years, we may surmise that the latter had died in the meantime, and of the 46 that stepped into their places, at least one third were blessed with a fair share of temporal riches. The number of journeymen was therefore probably between 20 and 30, and as each journeyman had the right to employ an apprentice, and as many also took in students, we may reckon the number of persons engaged in the manufacture of violins at somewhere between 60 and 70.

Among these there were but 30 families of violin-makers represented, if one identifies the variously-spelt names of Pöpel and Schönfelder. Of the house of Reichel there were ten master-workmen, of that of Schönfelder eight, from the families of Popel, Hopf, and Dörffler respectively, three each, so that the members of these five families alone nearly equalled the remaining 25 in numerical strength. Of the 14 families mentioned in the municipal registers of the year 1627 we only find the names of Brann, Glier, Pollman, and Jäger, in addition to the before-mentioned family names. We must not lose sight of the fact that in the early history of the Guild, the addition of new families of makers to the industries originally founded by immigrants was very slow. But we should not lay too much stress on this, lest it should be thought that the new comers and the older and respected citizens held aloof from each other, for it is an undoubted fact that the original 10 exiled families took especial pains to live on terms of close unity with the inhabitants of Markneukirchen.

Were not all in this land, where the very cradle of the Reformation had stood, and in a town whose church had been consecrated by the Evangelical services held therein by the great Reformer himself, bound together by the ties of an identical religious belief?

This meant a great deal more in those times, when State and society, and all the vital interests of a mixed population depended on the security and freedom of religious practice, than one is accustomed to realize in the present century of overwhelming enlightenment, of religious half-heartedness and of individual free-thought, if one may so express it.

We shall be probably correct in surmising that a long time before the formation of the Guild, violins were made in Markneukirchen by the founders of the Guild themselves and by their immediate ancestors, and that those seven original families (or at least the more important among them) had been settled in Markneukirchen a hundred years before, and that this lapse of time had amply sufficed to convert them from exiles into full-blown citizens.

The first charter of this Guild of violin-makers was granted and confirmed by the Herzog Moritz of Sachsen on March 6th, 1617. We must remember that the persecution of the Evangelists in Bohemia did not begin in the 17th century, but in the days of Kaiser Rudolf II. (the pupil and patron of the Jesuits), whose reign began in the year 1576, and Bohemia only succeeded in the year 1669 in obtaining the Royal edict granting freedom to the practice of Evangelical religious belief, which freedom, as history shows us, did not last long.

A second no less weighty reason for our belief in an earlier commencement of the pursuit of violin-making in, and the unseverable affiliation of, the violin-maker families to our city, lies in the conditions necessarily precedent to the pursuit of the craft, and in the habits of life of that period itself. The possession of a house, and at the same time the acquisition of rights of citizenship laid at that time upon the owners of such privileges the necessarily primary and frugal groundwork of a thriving industry. The violin-making handicraft was distinguished from simple handiwork (whose practitioners delivered their wares on the spot for immediate use, and, as a rule had not far to go for materials) not only by the difficulty of disposing of its manufactures, but also by the great care and far-seeing calculations which come into requisition in the procuring of seasoned woods, of strings, and other materials necessary to the completion of the instruments.

Such an art could only yield a certain (or indeed any) profit to those who had connections with districts and countries where the same art was already flourishing, and who knew the markets for the obtaining of its materials, and the sale of its products. The requirements of family life demanded, even of the skilled, that other means of livelihood such as beer-brewing, cattle-breeding, building, inn-keeping, and such other trades as the locality offered, should not be neglected.

But the living and work-rooms in the master violin-maker's house, not devoted to his mere trade, were entirely at his disposal, as they chanced to suit him for the warehousing of materials and perfected instruments, and for the composition of varnish.

Taking, therefore, these circumstances into consideration, the non-ownership of a house by a postulant for the mastership appears to have been regarded as a social stigma and disqualification for membership of the guild in many of the recorded transactions and deliberations of the assembled masters.

The banquet given on his admission by every new master, at which all the existing master-workmen and their wives were in the habit of being present, could not take place in the apartments of a new master who was not a householder. A person so situated had to give a considerable money compensation in place of the customary banquet, in spite of which compensation his position in the guild remained an inferior one. Thus we find that the two brothers (or cousins), Reichel and Caspar Schönfelder, possessed each of them his house, and were duly-qualified burghers of Markneukirchen.

It is recorded of Johann Reichel (Christian Reichel's son), who, soon after his father's death, was admitted master (in the year 1697), that he took a house and proposed to set up housekeeping with his mother. Johann Caspar Reichel was next-door neighbour to Caspar Schönfelder, who, besides being a master violin-maker, was also a baker by trade.

Though Caspar Reichel's son Christian, and his cousin Johann, were admitted masters in 1697, on the same day, the two must have pursued their trades in separate houses, as it would have been highly unseemly for two, or, worse still, for three masters of one family, to live with their families in one house, where in addition room had also to be found for a widow of the family. One Reichel house was situated at the epoch of the foundation of the Guild, in the middle of the market place, where it stood until the year 1840, while the other, so far as the available sources of information throw light upon the subject, was beyond it in the market place, next to the Burgher Schönfelder's house. As far as the families of the remaining five founders are concerned there is not sufficient reference in the Book of the Guild to prove their possession of burgher houses, yet there is little doubt that such proof could be found elsewhere.

In such cases at a later period, as make no specific mention of the want of a house-holder's qualification on the admission of any person to mastership, we may accept without hesitation the inference that the admitted master was a burgher or house-holder.

The designation of "Exiles from Gras-litz" must not convey the idea that these exiles had only just settled down, bought burgher houses, and immediately started the Guild of Violin-makers in Markneukirchen. Just as in Prussia the French refugees, and in the next generation the Salzburghers, were harboured by the Hohenzollerns (ever the protectors of Protestantism) and retained the designation of "Refugees" amid their new surroundings; so in Markneukirchen the children and grandchildren of the first exiles continued to be known as such.

The second half of the first century of the history of violin-making in Marknenkirchen, which, alas! the Book of the Guild only allows us to follow up to the year 1772, shows unmistakeable signs of soaring ambition, of a great consolidation of the trade and of its statutory ordinances, and a greater influx of new masters, although the increase in the number of established families living by the art was comparatively small.

It has been already shown by the list which we have given that of the non-exiles who had settled there in the 16th century the old families Brann, Götz (Goetzel), Pohlmann, and Glier, had been taken in; we need only add besides that the exile-family Papel (Pöpel) is casually mentioned in Markneukirchen in 1632.

The following are the masterships conferred between the years 1727 and 1772 to which the numbers denoting new families are added in parenthesis:—

1. Friedrich Pöpel, 1727, who was apprenticed to his father.

2. Johann Adam Schönfelder, son of the deceased Burgher and violin-maker, Johann Schönfelder, 1729.

3. (I.) Johann Christoph Schetelich, who married a master's daughter, 1729.

4. (II.) Hans Christian Uebel, who probably did likewise, since he, like the last-named master, was admitted on payment of

the half tax of 6 thaler and 18 groschen, 1729.

5. Georg Heinrich Kurzendörfer, 1729.

6. Johann Elias Pfretschner, who paid the dues claimed from a master's son, and was therefore probably son of Johann Adam Pfretschner, who was admitted in 1716.

7. Adam Voigt, the youngest son of a master (see No. 27 in the former list).

8. Johann Georg Lorenz (see No. 38 in the former list)

9. (III.) Christian Friedrich Meinel was admitted on payment of 8 thaler, exclusive of other dues, because he married a master's daughter, 1730. The family is alluded to as domiciled in Markneukirchen in 1632.

10. (IV.) Hans Adam Wurlitzer, who "intended" to marry a master's daughter, and therefore was admitted on the condition of a possible subsequent raising of the dues in case he should not fulfil his intention, 1732.

11. Johann Adam Kurzendörfer, for payment of whose dues Master Christoph Göze, citizen and clothmaker, went bail, 1732.

On this admission Conrad Adam Schönfelder and Christianus Reichel figure as headmasters, and from henceforward it became the rule that only the holder of the post of "Umpire" (adjudicator?) who was deputed by the council, should sign the protocol of admission.

This post through many generations was in the hands of the family Kretzschmar. The first to sign himself "Umpire" was Cornelius Kretzschmar, in the year 1710. Afterwards, from the year 1729 till 1772, the post was filled by Johann and Johann Gottfried Kretzschmar, probably father and son, the latter of whom was still living in 1779, after the peace of Teschen.

The father seems to have died in 1760, between the 29th May, on which date he signed the protocol as Umpire for the last time, and December 2nd, when we find Johann Gottfried Kretzschmar acting in that capacity.

In the year 1782 two new masters were created in addition to the above-mentioned, namely:—

12. Simon Voigt, a master's son, and

13. (V.) Gottfried Pilz, late musketeer, who taught himself the art of violin-making, and who, in consideration of whose service in the Kurfürstlich Horse, was admitted on easy terms. His admission was contemporary with that of Simon Pöhlmann, a citizen and " Defensioner " of the year 1696. Meanwhile, in the year 1734, we find admitted to Mastership :—

14. Johann Adam Reichel, second son of the head-master, Johann Reichel, the elder.

15. (VI.) Cark Wilhelm Heber, of whose origin nothing further is mentioned.

16. Johann Conrad Reichel, third son of Johann Reichel, the elder, 1737.

17. Christian Reichel, son of Master Johann Caspar Reichel, 1738.

18. Hans Adam Kretzschmann, born at Wohlhausen, 1738. He paid 30 thaler and gave besides a tun of beer. As the fees of admission to Masterhood for a stranger were generally about 12 thalers, the cost of the Master Banquet, without the beer, was pro-

bably valued at 18 thalers. Whether Hans Adam Kretzschmann was a relation of Georg Karl Kretzschmann (see former list) must remain uncertain, but it is not improbable.

19. Johann Adam Pfretschner, a master's son, 1738.

20. Hans Georg Reichel. fourth son of master Johann Reichel the elder, 1738.

21. Hans Georg Kretzschmann, who " intended" to marry the youngest daughter of Hans Martin Schönfelder, citizen and violin-maker, 1739.

22. Johann Georg Reichel, eldest son of Master Peter Reichel, citizen and violin-maker, 1740. A master, Johann Blech-schmidt, went bail for the payment of his admission fees.

23. (VII.) Hans Georg Martin, citizen and violin-maker's apprentice, 1742. He paid 22 thaler 5 groschen.

24. Georg Adam Gütter, youngest son of the head-master, Johann Gütter, 1743. On his admission Johann Caspar Reichel and Conrad Adam Schönfelder again sign as head-masters.

25. Johann Adam Klemm, citizen, son of Johann Georg Klemm, 1743. He, as well as his father, undertook peg-turning as a probationary post, and when his turn came filled the place of junior master of the Guild.

26. Johann Friedrich Reichel, youngest son of master Peter Reichel, citizen and violin-maker, 1745.

27. Johann Wilhelm Götzel, violin-maker's apprentice, from Wernitzgrün, son of master Johann Conrad Götzel, violin-maker and inhabitant of Wernitzgrün, 1747.

28. VIII. Gotthilf Fischer, violin maker's apprentice and a military man, in consideration of whose standing as a soldier, the taxes were reduced, 1748.

29. Johann Adam Gütter, son of the Head-Master, Johann Gütter, like the foregoing engaged in the Militia, and who obtained the consent of his Captain, 1748.

30. IX. Johann Christoph Schemmerling, a violin maker's apprentice, born in Dörpersdorf, who married the eldest daughter of the Master Gottfried Piltzen, 1748.

31. X. Georg Schuster, citizen and violin maker's apprentice, at whose admission the greatest caution was observed, for we find that the three petitions for admission, presented to the Guild since June 27th, 1748, were remitted for consideration from quarter to quarter, and full admission only followed on payment of 25 thaler, on the 10th April, 1749.

32—34. Johann Christian Hamig (Hammig), Johann Gottfried Schetelig, and Johann Gottfried Reichel, all three sons of Masters hitherto working as apprentices, 1751.

35. Johann Gottfried . Kretschmann, citizen and journeyman, son of Master Georg Carl Kretschmann, 1751. On this admission for the first time was required a vow to live " in accordance with the all-merciful mandate for the abolition of abuses among the workmen."

36 —38. Johann Freidrich Voigt, Johann Gottferid Pfretschner, and Johann Adam Fücker, violin maker's apprentices from abovementioned families, 1751.

39. Carl Friedrich Hamm, son of the ·citizen and violin maker, Johann Andreas Hamm, 1751.

40. XI. Adam Rossbach, violin maker's apprentice, son of master Adam Rossbach, ·citizen and saw-maker, 1752. He paid 29 thaler down for the master banquet and other dues.

41—43. Johann Georg Glier, Johann Friedrich Glier, and Johann Adam Fischer, violin makers' apprentices, sons of families above-mentioned, 1752. Johann Friedrich ·Glier paid 31 thaler because he was not a master's son, and Johann Adam Fischer 29 thaler for the master banquet, whereas Johann Georg Glier only paid 8 thaler 6 groschen, being a master's son.

44—46. Johann Georg Ficker (XII.), Carl Friedrich Liebel and Hans Georg Piltz, three violin makers' apprentices, 1752. The Liebel family was already mentioned in the roll of enfranchisement of 1627, in which Johann Liebel, is alluded to as council friend (? Raths-freund), 1752.

47. Johann Michel Seydel, the younger son of Johann Michael Seidel (*sic*), 1753.

48. Christian Gottfried Schonfelder, son of Georg Simon Schonfelder, citizen and violin-maker, 1755.

49 and 50. Johann Christian Ficker and Friedrich Wilhelm Meinel. the former, son of the Head-master Johann Caspar Ficker, the latter, son of the master, Christian Friedrich Meinel, 1755.

51. Johann Georg Kaessler, son of a ·citizen and violin-maker, 1756.

52 and 53. XIII. Johann Gottfried Lippold and Curt Friedrich Kretschmar, violin-makers' apprentices, 1759. The latter was descended from the Umpire family, and in consideration of his father and grandfather having for a long time occupied the post of Umpire (or adjudicator) to the Guild, he was charged smaller admission fees than other master's sons.

54—60. In the year 1760, on May 28th, no less than seven violin-makers' apprentices were given the rights of mastership at a single meeting. These are : — Johann Christian Voigt, Johann Conrad Reichel, Johann Georg Lippoldt, Johann Carl Pfretschner, Georg Adam Voigt, Johann Adam Martin and Johann Georg Kurzendorfer, all members of families already mentioned. Johann Adam Martin (not a son of the above-named Martin, 23) paid as a stranger 24 thaler for the right of mastership, whereas the other six, being masters' sons, only paid 8 thaler 6 groschen each.

61. The admission of Johann Adam Glier at the same assembly was a little more complicated. He was introduced into the Guild by his father, Johann Adam Glier, who also paid the first dues for him. He paid 5 groschen for fees and 1 groschen for duty. The formal admission was only completed on December 2nd of the same year on further payment of 7 thaler 12 groschen tax for the right of mastership (exclusive of the fees already paid). It was decided at the same meeting that in future all master's sons should be subjected to the same rules of admission and treated in the same manner as Johann Adam Glier.

62. Johann Georg Fücker was admitted to the rights of mastership on May 29th, 1760, and gave of his own free will half a barrel of beer to the Guild of Violinmakers. Fücker had been enlisted as a recruit during the seven years' war, when, by order of the King of Prussia and Kurfürst of Brandenbury (*i.e.*, Frederick the Great), recruits had been levied for the 10th Saxon Infantry Regiment. As he had found an opportunity of leaving the Prussian service and of entering the Saxon Service, he begged for the rights of mastership to be conferred on him. The full taxes were not claimed, probably on account of his being a soldier.

63. XIV. Johann Adam Nürnberger, 1761. He learnt the violin-making trade in Klingenthal from Johann Christian Eubel, won the rights of citizenship at Markneukirchen, and had half a mind to marry the youngest daughter of master Johann Reichel the elder. On this consideration the fees on admission to mastership were reduced to 10 thaler 16 groschen. In the event of his neither marrying the lady under consideration, nor any other master's daughter, the sum of 31 thaler was assessed upon him as a deferred payment.

64. Hans Georg Goetzel, from Wernitzgrün, son of Hans Conrad Goetzel, member of this Guild and an inhabitant of Wernitzgrün, 1761.

65. Hans Georg Reichel, son of Johann Conrad Reichel (see No. 16), and

66. Hans Georg Kretschmann (son of Hans Adam Kretschmann), both violinmakers' apprentices, 1762. They were

admitted on immediate payment of the mastership fees, and both gave besides 2 groschen 6 pf. towards the funds of the masters' Treasury. Reichel 12 groschen in addition towards the Exchequer of the apprentices.

67. Johann Friedrich Reichel, son of a deceased Johann Reichel, violin-maker's apprentice, was admitted as a member to this Guild, after compliance with the prescribed formalities, in 1762, in public session of the assembled Worshipful Guild of Violin-makers. He paid, besides the usual master taxes, 12 groschen to the masters' Exchequer and to the apprentices for pipes and tobacco. He was simultaneously admitted to participate in all the benefits which were enjoyed by other masters of the craft.

The records in the Chronicle concerning this admission are characteristic of the changed, or at least wavering, nature of the violin-maker's calling. These 67 masters admitted since 1727 had developed, together with a few dating from the earlier times, into a numerically important community during the past generation. The manufacture could now be carried on on a larger scale on account of the newer directions in which musical taste had progressed and the important development of concert music. But from these very causes the artistic aspirations of many of the makers became dulled into a commercial activity and the appreciation of the meaning of their art pursued *as* an art became more and more a thing of the past.

The descendants of fathers and grand-fathers, who had done splendid work under very difficult circumstances, felt satisfied with inferior work, restricted as they were in their efforts by the rules of a mere trades-union. By degrees they cut themselves off more and more from all connection with all matters foreign to the objects of the Guild, and the marrying of a master's daughter became the only means open to a so-called "stranger" of obtaining admission to this great community, conspicuous for its wealth and influence in matters municipal. It may be taken as a sign of the greater importance attached to household tenure that single members of the violin-makers' families devoted themselves to University studies, whilst outsiders, such as the two subsequently admitted pastors, Schetelig and the sons of the magistrates Kretschmar and Siebel, sought admission to the rights of mastership in the Guild. As early as in 1682 there had been some talk of a confirmation of the Guild by charter, but indications of official incorporation are first found in a proctocol of the year 1729, immediately following on one bearing only the theretofore usual designation of the Guild's work as a "worshipful" art.

The seven years' war was ended, and the arts of peace, especially music, flourished once more, and this renaissance brought with it higher efforts of mechanical skill, and a greater appreciation of the value of more delicate work, in the fabrication of musical instruments. This is noticeable already in the admission of masters in the following years.

68 and 69. In the year 1764 the violin-maker's apprentices, Johann Gottfried Hamm

and Johann Gottlob Ficker, were admitted masters in the Corporation of this worshipful art as before, and were only called upon to promise to conform to the rules of the community and the " Beneficent Mandates relating to the Suppression of Abuses among the Working Classes."

These mandates were first mentioned in the year 1745. As their title denotes, they were by no means restricted to the violin-maker's Guild alone, and indeed were only recommended to its notice because according to the artistic comprehension of that time, violin-makers were classed with the common workmen.

Notwithstanding these circumstances, however, it was an epoch in which the authority of the Corporation, with its regulated ritual, its control of the market, and command of remunerative work (with which control not only rights and privileges but also burdens and duties were combined), must have been highly beneficial in its workings, not only to the members of the Corporation and all who were in close relation with it, such as the apprentices, students, and journeymen, but also to all who stood without its immediate influence, but whose interests were in many ways connected with the Corporation.

What would have become of a branch of industry such as the art and craft of violin-making, in which the sale of the manufactured article depended entirely upon the absolute confidence which the buyer had in the gradually-acquired experience, and trade secrets of the maker, a confidence

which any superficial treatment or insuffi-
ciently clever imitation was calculated to
shape?

It being well understood that all within
the privileged Guild knew how to distinguish
between generally artistic and common work,
and that the Guild only offered as first-class
work, instruments which they knew to have
been wrought with the greatest care and
skill, combined with the use of best materials
only, not only were the interests of the
Guild served, but also of those of the
buyers.

Under these circumstances, the manufacture of musical instruments was in a position
to flourish and make strides. Had free trade,
as it has existed among us, for some decennia
past, existed at the time in which we write
it would undoubtedly have rendered such a
branch of industry as that of the manufacture of stringed instruments impossible as a
means of livelihood. Whilst the good com-
radeship existing in the Guild made it easier
for beginners in the art to obtain a trust-
worthy education in the requirements of
their calling, it certainly made the acquisition of the art cheaper, simpler, and more
likely to be tried with success by the isolated
member than if he had gone his own solitary
way independently. Apart, however, from
the few delicate nuances and superficial fads of construction such as are
and must be customary in the highly-
developed art of violin - making [as it
was practised in Cremona] they had to keep
on the whole to the fundamentary rules of
construction laid down by the great inventor,

Tieffenbrugger (*sic.*), who flourished in the year 1510.

From the beginning they were probably obliged to use for the better class of instruments, old and seasoned maple wood from Bavaria, the Tyrol, and other parts near to the Italian frontier, and, without doubt, they were also early in possession of valuable Italian instruments, which, once. recognised and used as models, were accessible to all the artists and craftsmen in the town.

In all probability, a few apprentices early found their way, during their wanderings, to Italy and other centres of instrument manufacture, and thus new ideas and extended knowledge reached the workmen engaged in the home industry. The two Fickers mentioned by Friedrich Niederheitmann in his "Charakteristik der Italienischen Geigenbauer"(Leipzig, 1877), point back to the first of the family of that name, a family still inhabiting Markneukirchen, and which, as we have already seen, has produced several masters.

Niederheitmann points out that the violins made by Johann Ficker in the years between 1719 and 1738 in Cremona have nothing in common with the peculiarities of the great Italian masters, of whose school Ficker was reckoned a member, but rather display an original technical intelligence. This traditional spirit of original art makes itself felt to this day in the studios of Markneukirchen, where violins of genuine German make are constructed, and sold at high prices, while in the other factories are turned out mere imitations of the most celebrated Italian masters, such as Amati, Stradivarius, Maggini. and Guarnerius.

It appears, from the records found in the registers of the Guild, that, on the admission of the later masters, the rule of the Charter of the Guild requiring that the applicant should have passed two years as a journeyman was relaxed, and higher fees were demanded for the Guild Treasury in lieu of service. In this matter violin-making stood on a different footing to other trades, and such a dispensation was not only unproductive of harm to the art, but often, especially in the case of masters' sons, demonstrably advantageous. Sons of families whose head members undertook trade journeys, had always the opportunity thus afforded them of learning new things whereby they could enhance and improve the status of their handicraft. It was certainly different where new comers were concerned, men who had no connection with the older families and who were obliged to content themselves with their own imitations of existing models and the amount of manual dexterity which their instructors had taught them. In such a case it was necessary to obtain the assent of the Head Authority of the State (the Landgrave) before the candidate to mastership could forego the years of probation as a journeyman. An instance of this happened at the admission of Adam Braun in 1697, but with regard to the enforcement of the two years' probation they seem, later on, to have exercised great indulgence. In 1748 a return was made to the strict observance of the articles of the Guild's charter as far as the formalities of admission were concerned. Hitherto it had been frequently regarded as an unnecessary loss of time to let the three prescribed taxes be paid at different times. As these were all paid at once it was possible to admit an apprentice to mastership within the limits of one day. After the first coming into force of the Landgrave's edict concerning the reform of abuses, the facilities afforded by a speedier mode of payment was still allowed to Gotthilf Ficker, a member of a family not hitherto connected with the art, but who had been enlisted in the Militia.

On the application of the violin-maker's apprentice, Georg Schuster, on June 27th, 1748, it was decided that the taxes should be payable quarterly; thus Georg Schuster was really only elected to Mastership of the violin-maker's profession at the Guild Meeting of April 10th, 1849. As he had to pay 25 thaler, he certainly cannot have been a master's son, ncr in possession of a Burgher-House, yet the family of Schusters was probably settled in Markneukirchen as early as the middle of the 17th century, for in a diploma of the year 1666 there is mention of one " Rauner Schusterin," i.e., certainly the wife of a man named Schuster who had wandered thither from the village of Raun (or she might have been a born Rauner) and in the year 1683 is mentioned the joiner, George Schuster, as owner of the site for the new cemetery.

On the admission, in 1760, of John Adam Glier, (son of the master of the same name), and of Johann Georg Ficker, who had performed his military service, the father, Glier, appears to have attached much weight to the observance of the articles of the Guild Charter, especially to that part relating to periodical payments; the time over which the payments extended lasted, for his son, from the 18th May to December 2nd, and the resolution was taken that in the future all masters' sons should be similarly treated.

If on the following admissions the terminal times of payment are not set forth in the Guild Register, it does not prove that the resolve was broken. Four candidates for master's rights who presented themselves for admission in June, 1765, and paid their taxes, only attained the full estate of mastership in June and July, 1766. The same rigour was observed in the subsequent admissions, although the candidates were scions of masters' families.

Still more severe, and conducive to the maintenance of the esteem which the excellence of their instruments had won for the violin-makers of Marknenkirchen, was the rule which enforced the production of one masterpiece by way of diploma-work, from all, without distinction, who proposed to practise the art and craft as an independent master.

In the original confirmation of the Charter by the Landgrave the requirement of a masterpiece had not been particularised, in many cases they had been satisfied with payment of a certain sum in place of the testing of work by the assembled Guild. Now the rule came into operation, that after payment of the third tax, the commencement of Masterhood should date from the commencement of the Masterpiece, and 20 days were allowed over and above for its completion. They went still further in prescribing the quality of the diploma work required. Just as to-day, violins of different qualities of higher or lower value are in demand [and we should consider it as an artistic impossibility for a Joachim, an Ole Bull, or a Princess Dolgorouki to try to produce the indescribable fruits of their artistic knowledge and power from the same sort of instrument as that used by gipsies, village-musicians, and ballad-singers] so in those days it was an undoubted necessity for the welfare of the "arts and crafts" that the Masters should be able to manufacture wares of different values. In this way they almost attained in one century to a distribution of work, expecting from one Master violins of inferior manufacture, and from another the fabrication of finer instruments, and the diploma works were judged accordingly.

Still in March, 1766, they resolved to inquire of the Council how the articles relating to the diploma-work in the Landgrave's confirmation of the charter should be interpreted. The Council gave answer in the interest of the manufacture and the trade, that each candidate, for the rights of mastership, should be obliged to deliver two diploma works, one an inlaid and the other a plain violin ; and thus the manufacture of an ordinary violin no longer sufficed for the attainment of mastership. According to modifications introduced later on [resulting from the continuance of the same ideas] it was permissible to make two inlaid violins, of finer quality, as Johann Gottfried Kretschmann did in the year 1767, or else one inlaid violoncello such as Johann Georg Reichel (violin and bass-maker's apprentice, son of the Master Johann Adam Reichel) manufactured in 1771 as his diploma work. This was the first occasion that the making of basses (violoncellos) was recorded in the Register.

Johann Adam Reichel, who died in the year 1794, aged 80 (probably born about 1714), and who had married the violin-maker, Wolf Ehrhard Lippold's daughter, Eva Sophia, it curtly described as a bass-maker, and this art was carried on by his descendants in both branches, and the Ecclesiastical Registers designate them all as " Bass-makers." It can hardly be inferred, however, that the art of bass-making was more generally spread than this.

We have Johann Adam Reichel I., a direct descendant of the originator, founder, and head master of the Guild of Violin-makers mentioned in the first half century of the Craft Register ; though we regard him as the first who made bass-making his chief study, and at last even his only work, yet there is always a possibility that there may have been isolated cases of other masters of the same family, cr relations of the family seeking to rival the Reichels. The stigma of narrow-minded exclusiveness (*Guild egotism*) does not, as we have seen from many facts recorded above, rest upon the Reichel family. Artistic and technological records point, however, to the fact that bass-making in Markneukirchen was limited to these families.

The entire 16th century, and also the 17th, was far less favourable to the development of instrumental than to that of vocal music, as Von Wasielewski has observed in his history of instrumental music.

Contemporaneously with the older stringed instruments of the Viol-family (such as Luther used to play in the circle of his family), the violin became more and more popular as an accompaniment to the voice ; at length, however, as "Queen of Musical Instruments" (being the one which nearest approached the human voice in quality of tone), it completed its triumphal progress through Europe and beyond it, distancing all rivals.

Perhaps the Romanians and their near Alpine neighbours developed their love of guitars, lutes, cithera, and such like instruments earlier than the Middle and North Germans ; with us the deeper, more penetrating and fuller toned music such as the violin afforded, moved men's souls more easily and more lastingly. As long as no perfect piano existed (for the spinnet was yet too meagre and expensive) violin-playing and music were inseparable. Not only in the houses of music-loving citizens, but, above all, among the great masses of the populace, in seminaries, from learned colleges to the humblest village school, and in the churches, the violin was indispensable. The new evangelical congregational singing as it gained in meaning. through the works of the sacred-song writers of the 17th century, depended principally on the violin as its medium of interpretation, and chief accompaniment.

In Markneukirchen itself there was a great number of so-called " choir-assistants " who regularly, with their violins, accompanied the singing in the churches ; and indeed. it was not likely that the descendants of those

noble-minded Bohemian settlers, who had
become exiles for their faith's sake, should
forget to give their faithful services to the
evangelical church-service, and to do honour
to the congregation by the added beauty of
musical support. This habit was so deeply
rooted in the lives of this religious sect, that
a mighty convulsion was caused among them
when, in the beginning of this century
(through discordance among the choir-direc-
tors of the church), only three violinists were
allowed to take part in the service. This is
recorded by Heinrich Gläsel in his record of
noteworthy facts written in the Vogtlandisch
dialect.

Meanwhile in Europe, and especially in
Germany, but little orchestral music was
performed, excepting perhaps at the petty
courts of the Landgraves, and at a few particularly musical centres. The researches of
the musical archæologists of to-day have only
succeeded in placing the origin of public
string-orchestral concerts at a hundred years
back, reckoning from the present time. It is
probably within the knowledge of the whole
world what service was rendered to music in
this respect by Spohr, Marschner, Schubert,
and Meyerbeer; but when we try to go back
as far as the 17th or even the first half of the
18th centuries, we encounter, even among
those learned in musical history, a conviction
that we are groping in the dark, unable to
form an opinion of any distinctively characteristic type of concert music, at least as far
as Germany is concerned.

In Italy, from classic times the home of
art, and in France, the two countries which

took the lead in, and sounded the key-notes
of, art and culture during the 17th and the
first half of the 18th centuries, it was
earlier understood how to blend harmoniously
together different musical instruments. In
the south of France, the cradle of the trou-
badours of chivalric times, German musicians
and instrument-makers were at an even
earlier date welcome arrivals, but they may
very probably have wandered thither from
Bavaria, the Tyrol and other south German
countries.

In the Museum of Musical History in
Berlin may be seen a sort of bass instrument,
called a "luther bass," but connoisseurs doubt
whether this was made in Luther's time and not,
as is more likely, put together some time later.
Violin-making was, without doubt, earlier
carried to perfection by the old Italian
masters than by the Germans, as is proved
by the Italian name "Viola," a name by
which all similar string instruments were
designated up to the middle of the 16th
century, after which time the diminutive
form "Violino" (little viol) was first
generally adopted ; the Italian masters
certainly played a form of string quartet, earlier than we Germans, yet it is questionable
whether the "Contrabass" (also called the
"Big Bass Violin" or "Violone") held the
same position then as it holds in the string
quartet of the present day.

The German musical author, Michael
Prätorius, describes the different bow-instru-
ments in the second volume of his work
"Syntagma Musicæ" (which appeared in
1,619) and distinguishes them only by the

manner in which they were played; thus :—
1. Viole da gambe, which are held between
the legs; and 2, Viole da Braccio, which are
held on the arm; the latter are also called
"little Polish Violins." The big viola di
gamba (contrabasso di gamba, with six
strings) did not find much favour on account
of the incompleteness of its tuning-mech-
anism; otherwise, he is of opinion that it
matters little how one tunes one's violin or
viola, so long as it is tuned to pitch, and the
player can perform on it well. It would be
clearly seen that even our modern four-
stringed bass only gradually attained an
enhanced importance, when we consider that
for a long time even the four-stringed violin,
invented by Tieffenbrugger, and improved
upon by his Italian followers, was played
upon, tuned to very varying keys, and that
till 1680 one played only up to the A above
the cleff, later rising to the C or Ut, and that
real virtuosity, as we understand the term to-
day, only began with Locatelli in the middle
of the 18th century.

So long as there was no specifically
developed string-orchestra music in Germany,
at least as long as the real string quartet had
not found a wide field, the bass was perhaps
a desirable but still not indispensable instrument, and it certainly required a series of
progressive stages of development before the
bass was recognised as a concert instrument
independent of the violin—before it reached
such popularity that without the accompaniment of the violin it was played in
village music supported only by the bagpipe.
In later times its use and importance

depended mainly on the position which the all-conquering violin attained in the orchestra, where, at first, in the mixture of all attainable instruments, the deep-toned trumpets, the trombone, the bassoon, struggled for supremacy with the contra-bass as the ground-work of the harmonies. Among the " Gamba " instruments the most popular was certainly the small ones called " Celli " (Violoncelli), but it was apparent from the already quoted work of Prätorius that even in a band composed according to the canons of Italian taste, the Viole da Gambe were only used "when they were to be had," and in default, the Viole de Braccio (modern viola) were considered as sufficient substitutes.

The whole of the 17th century was an era very unfavourable to the development of orchestral music, owing to the unceasing din of war, and the consequently persistent abandonment of all higher culture and art. Meanwhile the strides that had been made in establishing the importance of the " Trompette Maggiore " (big trumpet), and the " Tromba da Tirarsi " (slide trumpet), by Girolamo Fantini, served as a foundation for still more firmly establishing the prestige of the violin. For, as a matter of fact, the violin attained its position as a ruler of all instruments, only at the time when the simplest form of trumpet had reached its zenith. It was only in the first half of the 18th century, when Johann Sebastian Bach (1750) had inaugurated the renaissance of church music, that the orchestral standing of the violin was also established. Bach was the first to make the artistic polyphonic treatment of musical sounds universal in Germany. In his oratorios, side by side with isolated wind instruments, the string instruments were provided with the most difficult and complicated solo parts, and it must be remembered that Bach was director of the Thomas School in Leipzig, whither a commercial high-road from Bohemia led through Markneukirchen. Our native instrument-makers and those of other towns were in the habit of mingling in intercourse here, and to the present day, there exists at Markneukirchen a musical confluence at which during fair times the whole of eastern Europe satisfies its demand for violins. After this has been established it is hardly open to the smallest doubt that if, as we have seen, the manufacture of basses only began in the 18th century in Markneukirchen, the incentive came from Leipzig and from the introduction of church-music, and that in this respect (*i.e.*, bass making) the family Reichel distinguished itself.

A new orchestral wave soon after permeated from the southward to our musical valley. Joseph Haydn, bandmaster to Prince Esterhazy in Vienna, became, between 1760 and 1790, founder of the modern orchestra, by the invention and introduction of the symphony, as opposed to opera, which had by that time come over from Italy. Haydn first founded the real quartet, in which each instrument of the viol family held its special place ; the treble, alto, tenor, and bass viols, and it was at this period that the bass became the real foundation of the whole scheme of orchestral harmony.

With this epoch-making progression in the
musical history of the German people the
bass-makers were obliged to keep pace, but
for the less expert members of the community
it remained a daring undertaking to give
themselves over exclusively to this branch of
manufacture. Owing to the continued com-
petition of Bohemian, Austrian and Bavarian
instrument-makers, only such makers or
dealers as commanded the entire field of the
violin-making art were intrusted with com-
missions, makers whose position was a
guarantee of the excellence of build and
acoustic properties of the instruments, and
this is a condition of things which, we see,
continues to the present day. In consequence
of the defective means of intercourse existing
in these times (the Thurn and Taxis Courier
service refusing to forward these delicate
instruments) the trade in basses was by no
means a simple or easy one, although they
were only required singly. From all appear-
ances, there appear to have been a number
of families even then who, by preference,
occupied themselves with the sale of finished
instruments, persons who undertook all the
difficulties incident to the transport of violins
and other small instruments such as wood-
horns and flutes (which were manufactured
during the 10 years immediately preceding
the close of the records we have under consideration) and even charged themselves with
the trade in, and transport of, basses. Isolated houses of the Reichel family may,
however, even earlier have devoted themselves to the business, merely dealing in the
instruments; and we may infer this from the

fact that Crasselt mentions in the beginning
of the next century, besides a violin-making
Reichel family, three or four families of
dealers or traders of the same name. Three
families named Paulus are also mentioned as
violin-makers. We must assume that this
family occupied itself with the other trades
prevalent among the citizens of Markneukirchen, such as cloth-making, and especially
string-making (in which calling we find them
employed till the end of the century), as
they are not mentioned in the register of the
Guild among violin-makers proper before
1772, yet at the very latest they must have
become inhabitants of our town in the
middle of the previous century, since at that
time they were already closely connected
with the Reichels by marriage. Anna Mar-
garethe, née Paulus, is mentioned in 1749 as
widow of the Master Christian Reichel. The
Paulus family was also connected with the
original colonist family of Reichels by ties of
religious sympathy as they were descended
originally, on the mother's side, from one of
the Protestant families driven, in the year
1640, from Austria on account of their faith,
and which settled first in the village of
Wohlbach, between Schöneck and Markneukirchen, whence they subsequently came
over to Markneukirchen. We learn from a
chronicle in the possession of Herr Richard
Schuster that an old man of the name of
Johann Georg Paulus died during the course
of a commercial tour in Paderborn, and left
among his personal effects two basses,
besides several small packets of string and
other appurtenances. Certainly this Johann

Georg Paulus did not begin his commercial
tours in his old age; he had probably dealt
in the basses made by his friends and relations, the Reichels, for decades. We may
take as an example of the difficulties with
which the fabrication of, and traffic in, basses
were beset, that a member of an old established family of citizen manufacturers, Carl
Friedrich Jacob (born in 1778), was designated as a bass-maker on his death in 1814,
whereas he had previously been by turns
carpenter, locksmith, violin and general
instrument maker.

The manual dexterity of the hill-folk is
properly looked upon as an important factor
in the history of modern industries, and in
the Vogtlandish hill country this aptitude
seems to have attained a particularly wide
development. It would otherwise be difficult
to understand how in an age like the 17th
and 18th centuries, in which the different
manufactures were monopolised by so many
different Guilds or Trades Unions, the same
person could be by turns (to judge by the
extant accounts of isolated individuals), shoe-
maker, tailor, furrier, cloth-maker, saw-smith,
and then string-maker. But even though
different branches of the musical instrument
making industry were frequently practised
by the same person, yet it may be inferred,
in view of the progress in the technics of
instrument-making, that it was less in consequence of a many sided aptitude for art work,
than in consequence of a mere endeavour to
find occupation (amid the many channels of
activity open to craftsmen employed in lucra-
tive and artistic trades), that craftsmen eked

out a livelihood by first one and then
another branch of the art and craft of instrument-making.

It would be a hopeless task to attempt to
trace the gradual progress that was made in
the selection and treatment of materials, in
the excellence of production, and in the
ornamentation and beauty of the finished
instruments. Hyacinth Abele, a thorough
savant in the history of the violin and its
construction, laid down as early as the year
1864 the axiom that the important point of
such an investigation would be to find out, if
it were possible, how far each master kept to
his own pattern, what deviations he allowed
himself, and what improvements he projected
and carried out.

Indeed the intellect which could succes-
sively master the difficulties incident to the
execution of this task (a task almost beyond
the bounds of practicability) might indeed be
called an enviable one. Anyhow the fact
remains that, in the establishment of every
proposed technical improvement, it was of
immense importance that the travelling
manufacturers and traders should strive to
educate themselves still further by contact
with musical artists by studying new models.
and by keen observation of every kind.

Travelling for purposes of trade as well as
of manufacture was indispensable at least in
the second half of the 18th century. In all
probability the practice of travelling then
came in again, the practice from which the
" journeymen " properly speaking took his
title before attaining the rank of master-
workman.

The admission of Christian Gottlob
Pfretzschner could only take place in July,
1766, on account of his absence abroad, but
he had applied for admission in June, 1765,
and had paid the preliminary tax. It can
hardly be doubted after this explanation and
those that have gone before, adverting to the
dangers to which such " travelling youths "
were exposed by reason of the war, that the
" years of travel " (as journeymen) and even
longer period of expatriation became more
frequent among apprentices. The Register
of the Guild gives no further information
respecting the direction in which such expedi-
tions were made, but it cannot be doubted,
on reference to other chronicles, that
Northern Germany, westwards and east-
wards as far as the Sclavonish countries, as
well as the south as far as Italy, were much
visited. In the historical course of things
Markneukirchen had become, to a certain
extent, a centre of the earlier and more
highly developed art of violin-making as practised in Italy and the countries of Southern
Europe, including Austria. The principal
depôts for the sale of the instruments must
have been found in Northern and Eastern
Germany and in the neighbouring Sclavonic
states.

When once the wholesale trade in musical
instruments was firmly established, the
artistic training of the apprentices and
masters was conducted less in the undertaking
of trade journeys than by the establishment
of new trade centres, and new connections for the further development of the
manufacture of and trade in the instru-

ments. It was also of importance that the travelling violin-maker should be on the look-out for gut adapted for the manufacture of strings, good and cheap resin, oils, colours, wood, shell, and other material for inlaying instruments, especially if he desired to become a dealer as well as a maker. It was more advantageous to procure these articles (especially the resins, foreign woods and gut) from Northern Germany *via* the Dutch harbours of Hamburg, than by the Mediterranean harbours (whence they had to make the difficult passage of the Alps), unless these materials were to be had at the fairs.

It is an established fact, and one within the recollection of men living on the spot at the beginning of this century, that in our town the manufacture of strings had begun in the year 1730, although it was not until the year 1870 or thereabouts that a charter of Incorporation was granted to the string-makers' guild.

It is further stated in this connection, that at first, only a few who had previously travelled abroad occupied themselves with this branch of the art and raised it gradually to its present status by laborious study, industry, thought, and expense. It is reserved for future and more minute research to discover the names of those men who first devoted themselves to this useful trade, and to find how far the development of this particular branch of industry may have influenced the history of violin-making. The wide pasture lands of the great North German low-lands favoured the keeping of large herds of sturdy native sheep, which furnished the gut necessary for string-making, gut as worthless in those as yet unmusical parts of the country, as it was useful and valuable in the skilful hands of the workmen of Markneukirchen.

In all probability, they only gradually penetrated from the west to the far-lying Ostsee territories, where a firmly established trade centre was discovered for the sale of the " Polish Violin," especially in Danzig, the trade metropolis of the still important state of Poland. But it is not likely that these " Polish Violins " were really invented in Poland, but only that they were sold there in great quantities.

A certain Johann Tängel, born in Danzig, settled here (Markneukirchen) in 1700, and devoted himself to violin-making until his death, which occurred at the age of 90 in 1757, and he is said to have taught the violin-makers of this town the compounding of a new varnish of fine quality. Of this Tängel there is no mention in the register of the Guild, and yet it cannot be doubted that he introduced into Markneukirchen a new method of preparing varnish.

Coming as he did, from the country that supplied Europe with amber, he probably understood the confection of amber varnish better than anybody else in Markneukirchen,. as also that of the red-brown varnish, at this epoch so popular in Italy. Probably Tängel, not being a member of the Guild, established an important trade between his friends in Danzig and the principal members of the Markneukirchen Violin-makers' Guild, so that posterity has preserved a grateful

remembrance of the Prussian, of whom
Crasselt has recorded that "he brought
violin-making to its highest condition of
prosperity."

Still, there were locally closer trade connections which were not neglected by the travelling artist. It may be remarked in this connection that the art of making forest-horns, and
wooden instruments derived its chief incentive from Leipzig and Hof, and that in the
latter branch of the trade a member of the
already mentioned Gütter family distinguished himself. It would over-step the proposed limits of this work to go further into
the details of the ever-varied demand for
violins on the one hand and for wind
instruments on the other.

It would not, however, be out of place to
remark at this point that the manufacture of
violin bows remained at first far behind that
of violins, and that for a long time bows were
made only by the inferior artisans.

The *Chronicle* relates that the art of bow-
making was introduced into Markneukirchen
by a citizen named Joseph Ströz. He died
in the year 1760, and probably had settled
there in the first decades of the century. We
may infer from this that an earlier connection
existed between our town and the centres of
musical industry in Bavaria, and that the
travelling apprentices and traders often dealt
there.

In the Bavaria of that day (for the
Frankish States abutting upon the river Main
were ruled by the Frankish branches of the
House of Hohenzollern and by ecclesiastical
princes) one can only reckon Mittenwald and

a few neighbouring places in Upper Bavaria as the centre of the musical instrument industry. At this place (Mittenwald) violin-making became established in 1683 by Egidius Klotz (the esteemed pupil of the celebrated violin-maker Stainer) some years before the proud Venetian merchants trans-ferred their business to Mittenwald on account of some offence taken at the yearly market at Botzen. In the Upper Bavarian and the neighbouring Tyrolese Alps where Jacob Stainer (who was a pupil of Amati) was born, the wood of the hazel fir was found which was so highly valued on account of the splendid tone it imparted to the instruments, and it did not grow in the Vogtland and the Fichtelgebirge. Besides this there was always something to be learnt from the masters in Mittenwald and Füssen, as they were in constant communication with the Italian masters, and especially with those of the Cremonese school.

In the beginning of the century there were in Vienna, Prague, Nurnberg, Würzberg, Bamberg, Langenfeld, and Franken, many violin makers, and it can hardly be doubted that the travelling makers from Markneukirchen frequently wandered thither, and even into Italy. The maple trees of the country round Markneukirchen and the surrounding territories no longer sufficed for the growing manufacture ; even at the beginning of the previous century it had been necessary to procure suitable maple wood from Bavaria and the Tyrol, where maple trees eight hundred years old were to be found. The south German and Austrian forests supplied

large quantities of pear wood, such as was
used for finger-boards, tail-pieces, and pegs.
Above all, for a long time it was necessary to
procure strings, ready made from Italy, as
indeed is the case to this day.

If the travelling dealers did not reach as
far as the centres of the string manufacturing
industry, such as Venice, Cremona, Treviso,
&c., they were obliged to satisfy their needs
either at the markets of Botzen (to which the
Venetians had retransferred their depôt after
a short absence) or at other instrument-
making centres.

It is not probable that the violin and bass
makers of the time chronicled by the Register
of the Guild had penetrated beyond northern
Italy in the south, and the Rhine Vistula in
the north; at any rate all record of such
wanderings is wanting.

Subjoined are the names of those masters
who were admitted from the year 1765 up to
the time when the register closes:—

70--73. Johann Adam Martin, Carl Friedrich
Pfretzschner, Christian Gottlob Pfretzschner
and Johann Gottfried Kretschmann, whose
admission was begun by payment of taxes
on June 9th, 1765, and was only completed
in July, 1766.

74—77. In the year 1768, Georg Adam
Kässler, Christian Gotthilf Fischer, Carl
Friedrich Reichel, and Johann Georg Voigt,
were admitted as masters, but Reichel's
third tax was not paid until the following
year.

78, 79. Johann Georg Schönfelder and
Johann Adam Kretschmann, two Militia
recruits, who were admitted free in 1769.

80. In the year 1770, Johann Georg Reichel (son of the before-mentioned violin and bass maker, Master Johann Adam Reichel) became master.

81. The Master-book closes with the record of the admission of the apprentice, Johann Gottfried Voigt, 10th June, 1772, third son of Master Simon Voigt. The closing words are "Soli Deo Gloria."

When we look back on the long list of newly-admitted masters in the comparatively short space of 45 years, we must at once be struck by the difference between this period and that of the first 50 years. Among those of the last half-century we count altogether 30 families, seven old and 23 new ; in the first only 15 new families, inclusive of the families of Goetzel's not enumerated after the year 1761, when Adam Nürnberger from Klingenthal (who was engaged to a daughter of the Reichels) was admitted, no new parent stock was added to the Guild. Those admissions recorded in 1752 and 1759, viz. : those of Carl Friedrich Siebel and Carl Friedrich Kretzschmar were merely those of sons of families who were either magistrates or priests. Of the seven founders' families those of Poller, Knopf, and Rudert, have entirely disappeared, three each of the Schönfelder and Kurzendorfer families and one of the Pöpel attained mastership, whereas the Reichels contributed 12 new masters in addition to the 10 admitted in the first 50 years of the Guild, next to them of the older families stand in eminence the Voigts with seven new masters, the Pfretzschners and the Kretzschmanns each with six and the Fickers with five.

The four original parent stocks already named produced 19 masters, and later, up to the year 1723, the four newly-added parent stocks produced 24, thus making 43 altogether, out-balancing the whole 24 other families, which together only produced 38 new masters. On the whole in spite of the almost doubled number of masters, only about the same number of families as before (30—32) were really actively engaged in the pursuit of the art. Besides the 45 families recorded in the Guild Register, others afterwards wandered thither, not counting the seven emigrant families of the year 1677. There were six from Klingenthal, two from Schöneck, one from Wohlhausen, one from Wernitzgrün, and one from Dörpersdorf. Only a comparatively small number out of the 45 original families, have become entirely extinct. Such has been the case (I am informed) with the families of Poller, Püpel and Rudert, who took part in the founding of the Guild. Descendants of the founder families Hopf and Schönfelder still live in Klingenthal and Wernitzgrun, as violinmakers. Of the rest, with few exceptions, it would be easy to find descendants (in the case of most of the families) in such numbers as could hardly be equalled for magnitude by the noble families of the lowlands, or the patrician houses of the oldest and most celebrated towns.

As far as riches, fame, and family standing were concerned the Markneukirchen violin-maker families at first could not attempt to measure themselves with the celebrated violin-maker families of Cremona and other Italian centres of artistic industry.

The unpretentious character of their domestic habits was consequent principally, however, on the lower state of the development of artistic culture in Germany as com-pared with that of the Romaic countries.

In the present century, under the favour-able auspices of the newly-arisen German Empire, not a few families have attained world-wide importance by the magnitude of their manufacturing and export establish-ments, whereas in earlier days such establish ments were mainly Italian. Besides the works of isolated families in the violin-maker's Guild, later on [in view of the sharper defini-tions erected between wholesale and retail manufacture and trade] several other out-ward influences came to bear, as would be seen if the steps in the commercial progress of single families could be minutely traced. Minute researches into the history of the string-makers would probably throw much light on the subject.

The most eminent of the citizens of the Markneukirchen of to-day (which can com-pare with the most prominent towns of Saxony and Germany in respect of its com-mercial prosperity) are a splendid example of the beautiful adage "Work is the orna-ment of the citizen, and blessing is the price of toil," not mechanical work merely, but artistic work in the service of the Ideal. of Religion, and of Art.

We are justified in directing special atten-tion to the family of Reichel, in view of the large number of masters (22), which that family boasted within the comparatively short space of a hundred years, whereas the family of Schönfelder, who reckoned originally twice as many as the Reichels, only produced half the number of masters.

We may regard the development of the Reichel family as typical of the whole group of manufacturing families, seeing that outwardly their ways of life were exactly alike, and that they and their families were strongly bound together by ties by Guild fellowship and of art. The Reichel family was also connected during the 18th century, with the old native families of Markneukirchen, such as those (not at that time members of the Violin-makers Guild) of Ardler, Heinel, Wild, Roth, and Stark, the family of Lippold (which had settled there at latest in 1683), and the older houses of Schuster and Voigt rather than with the newer houses of Gütter Paulus, Ebner, Müller, Blechschmidt, Nürnberger, and Lederer.

The connections between the families of Lippold, Paulus, Reichel, and Schuster, are of such frequent recurrence that many of those to-day bearing those names are not only connected by family intermarriage, but also by blood relationship.

Other prominent families besides Reichel's were certainly not exclusive in their dealings with their less well-to-do fellow artists and fellow citizens. Judging from the Register of the Guild, the Reichels showed a kindly spirit of fellowship towards the apprentices and students in their choice and election of Masters of the Guild.

If the Reichel family appears to-day considerably less prominent in Markneukirchen than formerly, it must not be forgotten that

after the great fire of 1840, the principal
branches of the family left the town with
their capital (not inconsiderable for those
times), and found a field for their artistic
work and spirit of trade in other lands.   In
1851 the brothers Reichel won the gold
medal in Tilsit for their gut strings manufactured there, and exhibited in London at the
first International Exhibition.   Far from
home, they still preserved a faithful attach-
ment to their fellow countrymen ; they were
instrumental in enlarging the trade connections of their native town with foreign countries, and in the increase of their possessions
kept pace with the most prominent families
of their native country, without thereby
losing any of their love for their fatherland,
and the newly risen German Empire.

Many services to the community are
ascribed to them by those yet living, services
which, according to more correct authorities,
they did not render, as for instance, the
tradition that they were the first to establish
the trade of Sheep-gut with Odessa and
Tashkend, a service really rendered by the
old family of Weller, and the family known
at the beginning of this century as that of
the Burgomaster Christian Jacob Durr-
schmidt.   These encomia upon their name,
and the recollections of some still living who
were contempories of their youth, prove the
popularity of the Reichel family and the high
esteem in which it was held.   Members of it
have settled not only in many European
countries, but latterly even in more remote
parts of the world, thus enlarging their sphere
of mercantile operations and doing much
service to the trade of the world.

May the many who bear this name, or who related by ties of blood or sympathy, remember, on looking back at the examples of their fathers (who, for their faith's sake, forsook their free citizenship and their earlier Austrian home), that happiness in life is not dependent alone upon the possession of worldly goods, but on the prizing of higher and more ideal interests ; may they furthermore strive after simplicity and nobility of manner and custom combined with artistic spirit and piety, mingling the softer disposition of the south with the energy of thought and will that characterises the north, and may they ever remain true to their Vogtlandish country, and, above all, to the faith of their fathers !

" To God alone the glory !"